BRITISH FOSSILS

Peter Doyle

SHIRE PUBLICATIONS

SHIRE PUBLICATIONS
Bloomsbury Publishing Plc

Kemp House, Chawley Park, Oxford OX2 9PH, UK
29 Earlsfort Terrace, Dublin 2, Ireland
1385 Broadway, 5th Floor, New York, NY 10018, USA
Email: shire@bloomsbury.com
www.shirebooks.co.uk

SHIRE is a trademark of Osprey Publishing Ltd

First published in Great Britain in 2008

Transferred to digital print on demand in 2014

© Peter Doyle, 2008

Peter Doyle has asserted his right under the Copyright,
Designs and Patents Act, 1988, to be identified as the author
of this book.

A CIP catalogue record for this book is available from the
British Library.

Shire Library no. 474
ISBN: 978 0 74780 686 8

Designed by Ken Vail Graphic Design, Cambridge, UK
Typeset in Perpetua and Gill Sans
Printed and bound in Great Britain

MIX
Paper from
responsible sources
FSC
www.fsc.org   FSC® C013604

The Woodland Trust
Osprey Publishing supports the Woodland Trust, the UK's
leading woodland conservation charity.

www.shirebooks.co.uk
To find out more about our authors and books visit our
website. Here you will find extracts, author interviews,
details of forthcoming events and the option to sign-up
for our newsletter.

COVER IMAGE
Jurassic ammonites (*Dactlylioceras*) from Whitby, Yorkshire.

TITLE PAGE IMAGE
Androgynoceras, an ammonite from the Lower Lias
(Lower Jurassic) of Dorset.

CONTENTS PAGE IMAGE
Cast of *Charnia masoni*, an Ediacaran found by schoolboy
Roger Mason in Charnwood Forest, Leicestershire.

ACKNOWLEDGEMENTS
With thanks to Florence Lowry for access to the fossil
collections in her care, Tony Waltham and Colin Prosser
for use of their illustrations, Julie and James Doyle for
their support, and Nick Wright, Sarah Hodder and
Russell Butcher at Shire for their work in bringing this
book to fruition.

Illustrations are acknowledged as follows: R. Spekking,
page 4; M. Zincova, page 6 (both); M. L. Nguyen, page 11
(right); Ballista, page 11 (left); C. Eekhout, page 12; Paul
Selden, page 13 (bottom); Tony Waltham, pages 20–3
inclusive; and Colin Prosser, page 30 (top). All other
images are my own.

CONTENTS

THE MEANING OF FOSSILS

FOSSILS, the tangible remains of once-living organisms, exert a fascination for people of all ages. From the largest dinosaur to the smallest planktonic micro-organism, the fact that rocks contain the evidence of life long dead – and that they can readily be found and collected – is a revelation to many. The purpose of this book is to provide a simple guide to the fossils of Britain: how they were formed, where they can be found, and how they can be identified. Using this book, readers should be in a position to create their own fossil collections, and thereby glean some idea of the nature of fossil life through geological time.

Fossils have only been truly recognised as the relics of ancient life since the 'age of reason', that time in the late seventeenth and early eighteenth centuries when the basic principles of modern science were laid down. For mid seventeenth-century enquirers such as Robert Plot of Oxford, fossils – resembling so closely the remains of living organisms – were simply thought to be 'sports of nature', tricks planted by God to test the faith of the individual. Enlightened minds were soon to realise that fossils actually *resembled* shells and bones because they *were* once living examples of shells and bones, modified subsequently through burial and mineralisation to become stony objects, devoid of life. That being said, there are some otherwise lifeless stones that coincidentally resemble shells or bones, or the traces of ferns or other plant life created by chemical action; these are known as pseudofossils today.

Writing in the fifteenth and early sixteenth centuries Leonardo da Vinci was perhaps the first to recognise the true nature of fossils, but his work on the subject was to remain unread until the nineteenth century. Independently, eighteenth-century scientists such as Robert Hooke, James Parkinson (discoverer of the degenerative disease named after him) and James Sowerby were to place the study of fossils on a sounder footing, creating the basis for the modern understanding of fossil life. From these distant origins, the subject of palaeontology (literally, the study of ancient life) has developed into a highly technical science: it plots, amongst other things, the evolution

Opposite:
The horseshoe crab *Mesolimnus*, preserved in exquisite detail in the Jurassic Solnhofenimestone of southern Germany. This limestone is an example of a *Lagerstätte*, a deposit with exceptional preservation.

Above:
Insects preserved
in Baltic Amber.

Above right:
A pseudofossil: not
a fossil at all.
Fern-like
impressions on
paving slabs from
India, actually a
product of the
precipitation of
manganese-rich
minerals.

of life on Earth; it reconstructs the body plan of long extinct (and often mysterious) organisms, and conjures up images of ancient ecologies divorced in time from our own world. The succession of life through geological time, driven by the pulse of evolution, also has a role in providing a means of placing rocks in relative date order, the same principle by which archaeologists use coins or pottery shards to determine the sequence of historical events in an excavation. All of these aspects of the science of palaeontology draw heavily on an understanding of fossil life in its many forms, plotting out for ancient worlds what we would call biodiversity today – the extent of life on Earth.

THE FOSSILISATION PROCESS

Fossils are by definition the remains of once-living organisms that have undergone a process of fossilisation. What that process consists of depends to a large extent on what the local environmental conditions are following the death of the animal or plant. These conditions can vary from desiccation, drying out the flesh on old bones, through to entombment in ancient ice, such as the famous nineteenth-century discoveries of long-frozen mammoths in Siberia. For the most part, however, burial is a particularly important part of the process, one that protects the would-be fossil from further decay and attack by wind, water and scavengers. The importance of burial for fossilisation is illustrated by the fact that it is much more common for marine organisms to be fossilised than those living on land. Why? Because the land-based

Fossil fishes from Antarctica. These delicate fish fossils were preserved by rapid burial in stagnant conditions 175 million years ago.

Fossils from southeast London: these shells were buried and preserved together in a lagoon that was forming in the area 55 million years ago.

environment is typically one that is constantly being eroded away by the action of wind and water. Just look at the headstones in your local graveyard to see that. There is a much greater possibility of organisms being buried in water-based environments than on land, as rivers eroding sediments and silts routinely dump their loads into the sea; these in turn settle out as the layers that will ultimately become rock over the course of hundreds, thousands, or millions of years.

Prior to burial, the remains of organisms are subject to erosion by wind or water, being buffeted along in rivers, or being broken up by scavengers. Once buried, however, these remains are relatively safe, save for the attentions of burrowing animals scavenging for food; or the actions of fluids passing through the pores of the sediments, and dissolving away the remains of shells and bones. For the most part, preservation is a chance affair, and full communities of organisms are rarely, if ever, preserved. Because of this, palaeontologists usually have to rely on other evidence to help them reconstruct past life. Trace fossils – the transient footprints and burrows left by organisms – are surprisingly common, and help fill in the gaps of our knowledge.

7

The fossil traces of an organism burrowing through sands 1.8 million years ago, preserved in the cliffs at Walton-on-the-Naze, Essex.

Fossils, then, can be any organism that has been buried, and has survived the ravages of time and the attentions of other organisms intent on breaking them up or eating their remains. Rapid burial is the best option, particularly where there is no free oxygen available to promote the decay of organic tissue through the attentions of bacteria. Here, in exceptional states of preservation that define what are now called *Lagerstätten* (fossil 'bonanzas'), such conditions

Shells of the bivalve *Polypecora* from Colwell Bay on the Isle of Wight, preserved with some of their original colour bands intact.

mean that soft parts are preserved, often in exquisite detail, providing palaeontologists with the privilege of glimpsing the past through a spectacular preservation window. *Lagerstätten* are few and far between, relatively speaking, scattered through the geological record and affording amazing insights into the geological past. The Burgess Shale of British Columbia and the Solnhofen Limestone of Bavaria are perhaps the most famous, but there are many others. Despite these bonanzas, our knowledge of the fossil record is still slim. As one famous palaeontologist put it, 'studying fossil communities is like studying the contents of a village graveyard – and then only after many visits by grave robbers.' This, then, is the complexity of fossilisation.

After burial, fossils may be preserved in one of many states. If conditions are right, hard parts may be preserved intact and unchanged; usually only in the rocks of youngest geological age, however. In some rare cases, this form of preservation may retain some of the original pigmentation – in shells, for example. More often, the minerals that make up shells and bones (usually calcium carbonate for shells, and calcium phosphate for bones) are gradually replaced by other minerals – silica (quartz), and pyrite (iron sulphide) are

Lower Jurassic ammonite *Eichioceras* from Charmouth, Dorset. The original shell material of this ammonite has been replaced by the mineral pyrite, commonly known by the name 'fool's gold'.

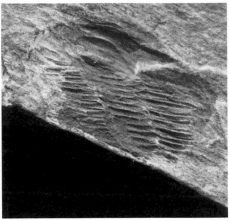

Above left: Logs of fossil 'monkey puzzle' tree (*Araucarioxylon*) preserved in the Triassic rocks of the Petrified Forest National Park, Arizona, USA. These logs have been preserved in three dimensions by silica seeping into their pores, creating the 'petrified wood'.

Above: Cambrian trilobite *Angelina sedgwickii* from North Wales. Examples of these fossils are usually found stretched and flattened due to the compressive forces that built the Snowdonian mountains. Scientists have had to work hard with a range of specimens to get a good idea of the exact proportions of this fossil.

Left: Mould of the interior of the bivalve shell *Myophorella* from the Jurassic Portland Stone of Dorset. The shell itself has been dissolved away, leaving impressions of its interior.

common, although other minerals might take their place. As these minerals will have replaced the intimate structure of the shell or bone, some remnants of its original structure might be left, visible through a microscope. Pore spaces may also be filled by one mineral or another, helping to keep the three-dimensional shape of the fossil intact and counteracting the natural pressures of the sedimentary layers above that would otherwise squash the fossil flat (a common feature of some sediments with fine grains, such as clays). Flattening and stretching are also encountered where the rocks themselves have been flattened and stretched, seen typically in mountainous regions, where the act of creating the mountains deformed the rocks. In porous sediments like sands, fossils such as shells or bones will be dissolved away leaving only moulds of the interior and exterior of the fossil. This will preserve the detail of muscle attachments on the interior, for example, or will preserve in negative the exterior ornament of shells, but all traces of the mineral structure will be lost. The moulds created may be filled subsequently by other minerals, but no evidence of the shell or bone structure will be seen in the resulting casts.

THE GEOLOGICAL RECORD

In order to get a handle on what was historically known as 'the succession of life through geological time', geologists and palaeontologists traditionally have divided up the record of the rocks into slices based on the types of fossils that they contain. This procedure was worked out in the closing years of the eighteenth century by two men, working independently – William Smith in England, and Baron Georges Cuvier in France. Recognising that fossils appear in layered rocks in a specific and predictable order, Smith was able to draw up a map of the distinctive strata – literally, the geological layers – of England and Wales, a map so significant that it has been described by one writer as 'the map that changed the world'. Cuvier, comparing fossil mammoth bones with those of the living elephant, was to discover that animals became extinct, never to appear on Earth again, and that, therefore, each species had its time. The mammoth was, as Cuvier was to infer, not only distinct from the elephant, but it was also long dead. Both Smith and Cuvier demonstrated that the geological record could be divided on the basis of its fossil content alone, and this principle is still used today.

Below left:
William Smith, a bust in Oxford University Museum of Natural History.

Below right:
Georges Cuvier, a bust in the Louvre, Paris.

The first fossils are known in some of the most ancient rocks on Earth – in Australia and Africa, where three-billion-year-old silica-rich rocks such as cherts (a type of flint) preserve, incredibly, the minute cells of rather unimpressive but nevertheless vastly significant organisms. These fossils, the first to be preserved on the planet, require specialist knowledge to unlock their secrets, and the same is true for many of the sparse fossils found in that part of the Earth's geological record known as the Precambrian (divided into the older Archaean, and younger Proterozoic), which stretches from the origin of the Earth some 4.6 billion years ago to the beginnings of abundant, recognisable life, some 540 million years ago. From this point on in geological time, anyone can recognise bones and seashells, corals and starfish, and the science becomes much simpler to understand. The UK – seen by many as the cradle of modern geology – is fortunate in having on its shores representatives of the most important fossil-bearing rocks. The oldest fossils in Britain belong to the last two thousand years of Earth history, with hard structures called stromatolites recording the growth of marine blue-green algae, and, at around 600 million years ago, the growth of strange quilted organisms belonging to a worldwide phenomenon – the Ediacaran organisms. These were first found in the late 1940s in the Ediacara hills of

Modern-day stromatolites growing in the shallow waters of Shark Bay, Australia. Such structures are built by blue-green algae, and their fossils stretch back deep into geological time .

southern Australia, and a homegrown example, the frond-like *Charnia masoni*, was discovered in Charnwood Forest, near Leicester, by schoolboy Roger Mason. Ediacarans have been found subsequently in Africa, North America and Asia.

Following the Precambrian came the Phanerozoic Eon (meaning 'evident life'), containing the fossil-bearing rocks of the last 540 million years. The rocks forming this record of time can be further subdivided, again on the basis of fossils, into three main units of time known as the eras Palaeozoic ('ancient life'), Mesozoic ('middle life'), Cenozoic ('modern life'). These roughly equate with what older popular science books have called the 'Age of Fishes', the 'Age of Reptiles' and the 'Age of Mammals'. Further subdivision is possible, down to layers of one-million-year duration, on the basis of more detailed examination of fossils. Since the turn of the twentieth century, these layers have been assigned ages in years, based on the study of radioactive elements preserved in the same rocks.

The interpretation of fossils as living organisms from all of these layers gets easier as the rocks get younger. Fossils of the Precambrian and Palaeozoic are distant in time, and it is difficult to find living organisms that might be directly compared with the fossil ones, in the manner of Cuvier and his comparison of mammoth with elephant. For the extinct trilobites of the early Palaeozoic Era, the horseshoe crab, common on the Atlantic Coast of North America, is the closest analogy; for the ammonites of the Mesozoic Era, the nautilus of the Indian Ocean is the best guide. For some organisms we may never know exactly how they functioned. In the Burgess Shale of British Columbia, a *Lagerstätte* formed around 510 million years ago, scientists have relied largely upon intuition to help them untangle the complex of organisms. The Burgess Shale contains familiar, hard-bodied trilobites, but a range of other organisms too, many unknown and bizarre. Celebrated is the soft-bodied animal, *Hallucigenia* (so named as it might have inhabited a bad dream), which, with its range of tube-like feet and spikes, was difficult for palaeontologists to understand. Giving it their best shot, they first reconstructed this horror one way up, then another, before settling on an

Nautilus pompilus, a shelled relative of the octopus. This living marine organism is the best guide to the life of the extinct ammonites.

Marella, the 'lace crab', a common fossil in the Burgess Shale of British Columbia.

13

animal with tube-like legs and an armoury of spikes on its back. Opinions vary on some of the most complex organisms.

NAMING FOSSIL SPECIES

Fossils are named like any other organism, as species, but the naming of fossil species is not straightforward. The identification of a new living animal species is based on two factors: first, the way it looks; and second, the fact that it lives with others of its own kind and reproduces to form young, which in turn form the basis of a new population of the same organism. For fossils, we only have the way it looks, its shape, form and similarities to others – we have little to help us work out whether or not it was able to breed successfully. In part, this explains why palaeontologists sometimes change their minds on the naming of fossils: while we can be sure about the breeding habits of living species, we have no such evidence about their fossil versions.

Like biologists, palaeontologists employ the system of naming that was first invented by the Swedish naturalist Karl Linnaeus in the eighteenth century. He suggested the use of a hierarchy of names, the species being the

Karl Linnaeus was the first to set up a workable system for the naming of all organisms, living or fossil. This is the first edition of his work.

Geological timescale			
EON	**ERA**	**PERIOD/SYSTEM**	**AGE (m.y. ago[1])**
PHANEROZOIC ('Evident Life')	CENOZOIC ('New Life' – 'Age of Mammals')	QUATERNARY	1.8–present day
		NEOGENE[2]	23–1.8
		PALAEOGENE[2]	65–23
End-Cretaceous extinction →	MESOZOIC ('Middle Life' – 'Age of Reptiles')	CRETACEOUS	145–65
		JURASSIC	199–145
		TRIASSIC	251–199
End-Permian extinction →	PALAEOZOIC ('Ancient life' – 'Age of Fishes')	PERMIAN	299–251
		CARBONIFEROUS[3]	359–299
		DEVONIAN	416–359
		SILURIAN	443–416
		ORDOVICIAN	488–443
		CAMBRIAN	542–488
ARCHAEAN & PROTEROZOIC (PRECAMBRIAN)[4] *(earliest fossil life known from rocks 3,500 million years old)*			4,600–542

Notes: [1] Time in years is determined from the radioactive decay of certain minerals preserved in the rocks. The exact dates used vary as science advances.

[2] The term 'Tertiary' has been used historically for these intervals of time.

[3] Americans often use 'Mississippian' and 'Pennsylvanian' in place of Carboniferous.

[4] The Precambrian is usually divided into the Archaean (4,600–2,500 m.y. ago), and Proterozoic (2,500–540 m.y. ago). It encompasses most of the Earth's history to date.

lowest level of the hierarchy, and the level that distinguishes most average living organisms. The Linnaean System employs two names to identify an organism: the species name itself; and the larger group that links it to similar species, the genus. Genera are then successfully nested into larger and larger groups, including families, orders, classes and so on. Latin and Greek are used for these names to provide an international language that can be used across the world. This makes the names seem old-fashioned, sometimes difficult to pronounce – but it allows scientists and amateurs alike to give formal names to their finds. Modern human beings, who mostly look alike and who can successfully interbreed, are encompassed in the species *sapiens*, and the genus *Homo*. Older, fossil, species of the genus *Homo* include our immediate ancestor, *Homo erectus*, and the family Homidae includes the genus *Homo* itself, and our more distant relative, the genus *Australopithecus*. All species, living or fossil, are grouped in a similar manner.

THE BRITISH ISLES – 2,900 MILLION YEARS OF EARTH HISTORY

Britain can claim to be the birthplace of modern geology. It was here that, at the turn of the eighteenth century, William Smith was to map out the distinctive geological layers – the strata – that provide very direct clues to the evolution of that small part of continental crust that the British call home. Smith was working during the burgeoning years of the Industrial Revolution, developing a network of canals as a civil engineer. His purpose in drawing up the map was to plot out the range of distinctive geological units he saw, deriving a knowledge of the rocks that would serve him well in considering the approaches to be taken in cutting and preparing canals. Importantly, Smith knew that while strata varied layer-on-layer, the distinctive fossils that they contained also changed. Geologists realised that layers of rock stacked one upon another represent a record of geological time, each layer recording the events of a particular time interval. Spectacular slices through the Earth, such as those of the Grand Canyon in Arizona, USA, cut down through successive ages that we now know date from around 170–1,840 million years ago. Through those millions of years life on Earth changed, and left a record of fossils in each layer that is distinctive. This was the principle (known as 'Faunal Succession') that was first established by Smith two hundred years ago, and which is still used today by geologists.

Since 1835 the British Geological Survey has taken on the task of mapping out in detail the various rocks and their contained fossils across Britain. The full map shows a lot more than simply the layering of rocks. The rocks seen by Smith formed largely when the land area of the British Isles was under water as seas rose and fell, a function of greenhouse periods of global warmth followed by icehouse periods of global cooling. The rocks also preserve the physical evidence of this small piece of Earth being rocked by the action of tectonic plates, activity that helped move Britain across the Equator from the Southern Hemisphere to its present position over 300 million years of Earth history. All of these changes have impacted on the nature of the rock layers formed, and on the nature of the animals and plants that were alive at the time.

Opposite:
The 'Organ Pipes',
columns of basalt
lava that form part
of the Giant's
Causeway World
Heritage Site in
Antrim.

the Chilterns and the Downs surrounding the younger Palaeogene (in Britain, the Eocene and younger layers of what was once known as the Tertiary) rocks that underlie London. A flexure (fold) in the rocks brings older rocks to the surface in the Weald of Kent (conveniently grouped under the title 'Wealden') but from there the magnificent chalk scenery of the South Downs reaches the sea at Beachy Head and the Seven Sisters, with the Palaeogene rocks (similar to those of London) being seen in Hampshire. All of these rocks are abundant in fossils, from the fabled ammonites of Lyme Regis and Whitby (a town with three of them in its coat of arms), to the sea urchins of the Chalk, and the clams and snails of the Palaeogene. Perhaps most importantly, it was from the Wealden sandstones and clays that some of the first dinosaurs in the world were discovered in the 1820s – a wholly British discovery overshadowed by the much later finds of the reptilian superstar *Tyrannosaurus rex* and his contemporaries in the American Midwest.

West of the Tees–Exe line are the red sandstones of the Triassic. Frustratingly barren of fossils (other than some very rare reptile footprints), these desert sands form a band that extends from the coast of Devon at

Above left: Lias (Lower Jurassic) cliffs of Lyme Regis, a favoured fossil-collecting destination for two hundred years.

Above: The Seven Sisters of Sussex, chalk cliffs south-east of the Tees–Exe line. The chalk, formed at a time of global 'greenhouse' warming, is rich in fossils.

Left: Triassic red sandstones north of the Jurassic outcrop, here forming the hill that houses Nottingham Castle. These sandstones are, frustratingly, largely devoid of fossils.

Sidmouth, along the line of the Severn, Trent, Ouse and Tees, and continues north-west into the Midlands, Cheshire and south Lancashire. Part of a great desert, these sandstones formed when Britain was a fraction of a huge 'supercontinent', a vast tract of land that encompassed almost all of the Earth's land surface, swept together by the action of tectonic plates moving across the globe. Geologists call this continent 'Pangaea'; the rest of the world was simply ocean. Having taken about 300 million years to construct, Pangaea would be wrenched apart over the succeeding 200 million years, to form the continents we know today. The Triassic desert was formed on a land surface that records the actions of earlier tectonic upheavals and, because of it, often the rocks do not lie as they were formed, on the horizontal.

South-west of the Exe the rocks of Devon and Cornwall belong typically to two groups: those formed by great river deltas during the Carboniferous (forming gritty, sandy rocks known traditionally as the 'Millstone Grit'), which are largely devoid of fossils; and the limestones of the 'Devonian' – rocks formed in warm tropical seas some 400 million years ago. Seen in the coastal town of Torquay, and in Plymouth, these hard rocks are useful for building, and contain a pleasing array of fossil corals and other more obscure fossil organisms. The South-west Peninsula is also famous for its granite – rocks formed from a molten state and forced into the Earth's crust – no fossils here.

WALES

Wales, and the Welsh borderland west of the Triassic red rocks (a distinctive change can be seen in the building stones of houses along the roads leading westwards), has a geological record that reaches back into the Cambrian, at least 500 million years ago. Slates and other rocks of Snowdonia, stretching southwards to central

The slates of North Wales contain fossils from the early part of the Palaeozoic. Slates like these from Nantile were quarried extensively for roofing.

Above:
Carboniferous
Limestone cliffs in
the Derwent Valley.
This limestone,
formed from
extensive tropical
reefs as Britain lay
close to the
Equator, is rich in
fossils, and helps
form the backbone
of England.

Above right:
The Millstone Grit
consists of coarse
sandstones used in
the production of
millstones, like
these at Stanage
Edge. The Grit was
formed in deltas
that spread across
northern Britain
before the
establishment of
the later
Carboniferous coal
swamps.

Wales and Pembrokeshire, and eastwards to the Welsh borders, belong to the earliest part of the Palaeozoic. Their names are of Welsh origin such as the Cambrian, Ordovician and Silurian. These rocks, first named by Roderick Murchison and Adam Sedgwick about a hundred and fifty years ago were the source of heated debate and discourse; the name Ordovician was invented as a compromise between the two men's beloved Silurian and Cambrian systems, by the diplomatic Charles Lapworth. (Tempers had been running high.)

All of these Welsh rocks contain fossils, frustratingly rare in places – and the tortured nature of these strata, caught up in successive mountain-building events, means that the fossils are sometimes difficult to detect in the older rocks. Trilobites, and the weird, saw-blade-like colonial fossils known as graptolites are the reward, amongst others. Southwards, Wales has a full suite of Carboniferous rocks; the Coal Measures of South Wales spawned the Welsh coal industry, now largely gone. Coal is almost a hundred per cent vegetable in origin, derived from the compaction of organic matter from lush forests, and the rocks that contain the coal (often the remains of the very soils that grew the trees) contain ferns and the remains of branches. Similar rocks are seen in the Clwydian hills to the north of the country.

NORTHERN ENGLAND

North of the Midland Triassic sandstone belt are the Pennines, a range of hills (the 'spine' of England) that are constructed from Carboniferous limestones and grits. The limestones, once called the Mountain Limestone for obvious reasons, are rich in fossils: sea animals that once thrived in warm, lush, tropical seas that seem at odds with the often bleak and windswept Pennine fells around Buxton, Castleton and the rest of the Peak District. The largely

fossil-barren grits and sandstones, part of what was once called the Millstone Grit (its stone ideal for the manufacture of millstones) accompany the limestones in constructing the magnificent scenery of the Pennines. West of the Pennines, the Lake District returns to the Ordovician and Silurian rocks, cousins of those farther south in Wales, exposed in mountainous terrain, and sadly largely unproductive for fossils.

SCOTLAND

Across the border, more productive are the Southern Uplands, a suite of Ordovician–Silurian rocks buckled and broken during earth movements that were to close an ocean (known to geologists as Iapetus) and unite England and Scotland for ever – part of the early days of the Earth's drive to create the continent of Pangaea. World famous for their trilobites and graptolites, the Southern Uplands still record the existence of this once thriving ocean. The prosperous Midland Valley of Scotland to the north returns to the familiar Carboniferous.

North of the Midland Valley, along a great fracture that geologists refer to as the Highland Boundary fault, there is a change in the geology of Britain. Gone are the fossil-rich limestones, shales and clays. In their place are hard, barren and crystalline rocks created when the mountains of Scotland were built hundreds of millions of years ago. Not all of Scotland is this way: rocks made from sediments are also to be seen, but as they stretch back in time to the Proterozoic and Archaean, before life became abundant, fossils are few and far between. The oldest rocks, in north-west Scotland, have an antiquity that stretches back 2,900 million years – the oldest rocks on Earth, from Greenland, being 4,600 million years old. Younger fossil-bearing rocks are seen

Above left:
Suilven, one of the mountains of north-west Scotland, composed of Precambrian red sandstones of the Torridonian. These early sedimentary rocks, formed under desert conditions, lie on top of much older crystalline rocks, devoid of fossils.

Above right:
The Old Man of Hoy, Caithness. The Old Man is composed of layered lime-rich mudrocks that split easily to make perfect paving stones. Formed in a lake that stretched to the Orkneys during the Devonian, these rocks are rich in early fossil fish.

scattered across Scotland, however; relics of seas that had the audacity to lap up against the highlands, mountains that may have once rivalled the Himalayas in altitude. The most famous belong to the Devonian Old Red Sandstone, once part of an earlier desert, and an earlier continent, than that which formed the Triassic 'New Red Sandstone'. The Old Red Sandstones of Caithness, Sutherland and the Orkneys were forming in a lake 350 million years ago – and preserve a spectacular array of early fishes in beautiful condition.

IRELAND

Ireland, Britain's constant companion during those 2,900 million years of Earth history, has a similar geological history. The rocks of Down and Armagh, Monaghan and Cavan, link directly with the Southern Uplands, and tell a similar story – with similar fossils of trilobites and graptolites. Opposing Wales across the Irish Sea, the Wicklow Mountains south of Dublin comprise rocks of similar vintage to North and South Wales from the Cambrian–Silurian, and record life as it was during this ancient period some 540–415 million years ago. Much of the rest of Ireland is blanketed with a thick covering of Devonian 'Old Red Sandstone' and Carboniferous limestones with its distinctive reef fossils, with the exception of more ancient Proterozoic rocks in the far north-west, and the famous lavas of Antrim. These were volcanic outpourings caused by the opening of the Atlantic Ocean some 60 million years ago that blanketed the older chalks of the Cretaceous period. In cooling, these lavas were to produce Ireland's most spectacular geological site, the Giant's Causeway, largely barren of fossil life.

THE YOUNGEST ROCKS

The youngest rocks of Britain and Ireland are those clays and sands that formed before and during the 'Ice Age', during the Quaternary, some 1.8 million years ago. The action of glaciers gouged out the Earth's surface creating deep wounds that allow us to see much of the older surface rocks. When the energy of these ice masses subsided, sediments were unceremoniously dumped, marking the end of the ice – ice that spread south to the outskirts of London. The most famous fossils associated with this time come from East Anglia, with the remains of Ice Age mammals being found in Essex (including, perhaps incongruously, animals we would consider to be more at home in the warmth of Africa, such as hippos and elephants), and falling from the crumbling cliffs of Norfolk. The Red Crag, coarse sands forming parts of the coasts of Essex and Suffolk, contain the youngest fossil marine organisms to be preserved on land, and confuse passers-by as their shelly remains mingle with modern-day shells on the beach.

The record of the rocks, preserving the evidence of the evolution of life on Earth, is well represented in the rocks of the British Isles, which, uniquely for

The 'Red Crag', distinctive red fossil-rich sands that form the cliffs at Walton-on-the-Naze in Essex, formed during the Pleistocene. These young rocks contain fossils of many species still alive today.

their size, preserve an almost full complement of the main geological time units within their shores. Studying this record in the early part of the nineteenth century, John Phillips, nephew of William Smith, was first to identify the three main eras of dominant fossil life, the Palaeozoic, Mesozoic and Cenozoic. Phillips based these distinctions on the rise and fall of biodiversity through time. He identified two low points, at the close of the Permian and the end of the Mesozoic. We now know these as two of the five main mass extinctions that affected life on Earth. The most pervasive of these is the Permian, separating Palaeozoic from Mesozoic life, a short interval of time that saw great changes of life in the shallow seas surrounding the supercontinent of Pangaea. Ninety per cent of all marine species were to be extinguished; the survivors were to spawn a new chapter of marine life. The second major dip in diversity seen by Phillips was at the close of the Cretaceous, well known to most people as the time of the demise of the dinosaurs and their cousins, as well as many other organisms besides. Other extinctions are known that were also to bite deep into the planet's stock of plants and animals, but none perhaps as profound as those at the end of the Permian and Cretaceous, the effects of which are well represented in the fossils of Britain.

THE VARIETY OF FOSSILS

THE museums of Britain and Ireland are stuffed with fossil remains that have been collected for centuries, some, like those in the Natural History Museum in London and the Sedgwick Museum in Cambridge, dating back to the early eighteenth century. These collections owe their origin to landed gentry who took time to build up 'cabinets of curiosities' for discussion in after-dinner soirées. Now, they provide scientists with reference specimens for continuing study, and browsing museum collections can be a rewarding experience for anyone interested in fossil life.

These collections reveal what we might expect, that those organisms easily observed alive today are those that we might easily observe as fossils in the past: plants, vertebrates (animals with backbones) and invertebrates (animals without backbones), with the last group being the commonest in the rocks of Britain. This chapter surveys the basic characteristics of the main fossil groups to be found; there are others, of course, particularly the many and varied organisms of minute size that can be found in many rocks, and grouped together as microfossils. Microfossils include everything from single-celled organisms to crustaceans (complex animals related to shrimps and lobsters). Requiring more preparation to see, and a microscope, these are not covered here. The next chapter covers some of the commonest fossils to be found in the UK.

PLANTS

The oldest organisms to be found on the planet are algae, fossil 'plants' dating back some 3,500 million years. Most geologists believe that these algae had a profound effect on the Earth's early atmosphere, changing it (by photosynthesis) from the volcanic gases of the birth of the planet, to something approaching the healthy mix we breathe today. It was the action of these early plants that was to secure the future of life on Earth, and to encourage the proliferation of organisms that have existed in the last 540 million years, the eon of 'evident life'. Plants as we know them, with tissue capable of passing fluids and nutrients throughout their stems, and growing

Opposite:
*Neohibolites
minimus*, common
belemnites from
the Lower
Cretaceous Gault
Clay.

Fossil ferns from
the Carboniferous
Coal Measures of
northern England.

on land, are known only from the Silurian, and rare examples have been recovered from Wales, now preserved in the National Museum in Cardiff.

From this early beginning grew the first forests, tied to swamps and damp lowland areas at first (for reproductive reasons), in the Devonian and Carboniferous. Here, horsetails, ferns and 'club mosses' grew into gigantic trees that created the coal forests of Britain and the rest of northern Europe, as well as North America. Their fossils can be found, often compressed, wherever coal was mined – sometimes spectacularly, with tree roots left behind *in situ*. Examples can be seen preserved in many parks (including, famously, Victoria Park in Glasgow). The development of enclosed seeds meant that plants could spread inland and, when dinosaurs roamed the land, so conifers, cycads and ginkgo trees grew on mountain foothills. Conifer wood, floating out to sea, is a common fossil in the UK. Flowering plants, another reproductive innovation, only bloomed at the end of the Cretaceous, just as the dinosaurs were in their death throes. With the flower came the rise of the insects and, from this, the spread of flowering plants across the globe. Fragile, relative to the shells and bones of animals, plants are nevertheless well represented in the fossil record of Britain.

VERTEBRATES

The evolution of animals with backbones treads a familiar course that has been described in many books since the discovery of their fossil remains in abundance in the early part of the nineteenth century: fishes to amphibians, amphibians to reptiles, reptiles to mammals. Many of the discoveries that led to these links were first made in Britain. The earliest fishes are known from rocks as far back as the Cambrian, and in Britain from the Silurian. The discovery that one mysterious group of miniature teeth-like fossils (known to palaeontologists as conodonts) was actually the complex jaw apparatus of an early vertebrate came from discoveries made in the Carboniferous rocks of Scotland. Many other fossil fishes are known from the Devonian rocks of Caithness. Scotland has also seen the discovery, in spectacular preservation, of what is reputedly the earliest known reptile (*Westlothiana lizziae*) from Bathgate in the outskirts of Edinburgh, some 340 million years old.

The rise of the dinosaurs was first recorded in British rocks, the earliest examples being described from teeth and other bones in southern England: *Megalosaurus* (1822) and *Iguanodon* (1824). These early discoveries were later swept together into the Dinosauria, a name first announced in Plymouth in 1841. Uniquely, these dinosaurs (and other large animals) were reconstructed in lifelike three-dimensional form in Crystal Palace Park in south-east London, where they may be seen to this day. From these early beginnings, later discoveries in America and elsewhere later in the nineteenth century would greatly add to our knowledge. Contemporaries of the dinosaurs, marine reptiles such as ichthyosaurs and plesiosaurs have been found intact since the early nineteenth century in the Jurassic rocks of Whitby, and, famously, Lyme Regis, where fossil collector Mary Anning supplied the scientific community with many of its most spectacular fossils. Replacing the dinosaurs after the great extinction event 65 million years ago, the remains of birds and mammals are also found in British rocks, with the Ice Age mammals providing the largest evidence of this.

Below left:
The dinosaur *Iguanodon* modelled in 1854 in Crystal Palace Park, south-east London.

Below:
One of the American dinosaur superstars from the American Museum of Natural History: *Allosaurus*.

The 'Dudley Bug' is part of the coat of arms of the town of Dudley. Here a trilobite is carved proudly in relief on a column in the Town Hall (Colin Prosser).

INVERTEBRATES

As invertebrates — animals without backbones — are (barring bacteria and viruses) the commonest organisms alive today, so they were in the geological past. With a fossil record that stretches back at least 600 million years in the UK, invertebrate fossils are mostly the remains of sea creatures that have been preserved in the sediments that accumulate under water. The main groups, commonly found in the rocks of Britain, are described below.

TRILOBITES

These are sea-dwelling organisms related to crabs, lobsters and insects, in that they have an external jointed skeleton with multiple jointed limbs. Now extinct, their nearest living relatives are the horseshoe crabs of the Atlantic seaboard of the USA. Their name comes from the fact that they can be divided in to three 'lobes' from left to right; they are also divisible into three parts from head to tail. Trilobites have a great diversity of form, with various spines, pits and knobs displayed upon their head and tail. Most possessed compound eyes, externally similar to those of their relatives, the insects, but others, such as *Trinucleus*, had no eyes at all. To protect themselves, many of these bug-like animals were able to roll up, tucking their jointed limbs inside their hard outer shell. Nevertheless, trilobites are often found in pieces, as, like all animals with an external skeleton, they had to shed it periodically to grow. Trilobites are found in rocks of Cambrian–Permian age, and are well known from the Welsh borderland and the Midlands town of Dudley — where one example, the 'Dudley Bug', has pride of place in the town's coat of arms.

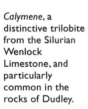

Calymene, a distinctive trilobite from the Silurian Wenlock Limestone, and particularly common in the rocks of Dudley.

GRAPTOLITES

These are somewhat mysterious organisms forming colonies that were mostly floating in the oceans at the same time as the trilobites were grubbing around on the sea floor. Resembling miniature saw blades, graptolites possessed tiny individuals that fed for the good of the colony, each housed in a little tube (each tube forming a 'tooth' of the 'saw blade'). Graptolite colonies may have had a single or multiple branches, which, in some examples, hung downwards from a float, or in others, spread upwards. Still others were spiral in form. These shapes helped graptolites to feed on nutrients suspended in the water. Extremely valuable to geologists, as these floating organisms were widespread and quick to evolve — and are thus good indicators of relative geological age — graptolites can nevertheless be difficult to find. Their name refers to the fact that they often look like pencil marks on rocks. They are found in rocks of Cambrian–Permian age, with examples well known from South Wales and the Southern Uplands of Scotland, as well as the Welsh Borderland.

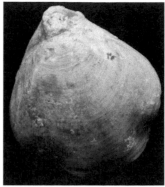

BRACHIOPODS

These are shellfish that were dominant in the seas of the Palaeozoic (Cambrian–Permian), but were replaced by the clams and snails after the great extinction event decimated much marine life (up to ninety per cent of all species) at the end of the Permian. We know a reasonable amount about how brachiopods lived from study of their living relatives. Today, we know these organisms fed differently from other shellfish, filtering out food particles using special 'arms' housed within their shells, supported on shelly 'brackets' that can be seen in some exceptionally preserved specimens. In the geological past, and today, most brachiopods attach themselves to rocks to feed, using a stalk that allows them to move around into currents in order to access the best food. Many examples have an opening through which that fleshy stalk (the pedicle) passed. Brachiopods were so common in the Palaeozoic that whole rocks are made up of their shells, and can be found all over Britain. Surviving the Permian extinction, brachiopods can also be found in much younger rocks; consequently, these unassuming fossils form a part of most people's collections.

Above left:
Silurian graptolites:
Monograptus.

Above:
Epithyris, a Jurassic brachiopod from Oxfordshire.

CORALS

These organisms have the advantage of a supportive, hard skeleton (made of calcium carbonate, the mineral substance that creates the rock limestone) and, as such, are often found forming fossil reefs. These reefs, although not identical to ones we would expect today, are sufficiently similar to allow us to make some educated guesses about their life and origin in the past. There are many types: the more ancient examples (rugose and tabulate types), which form the reefs seen in Silurian and Carboniferous limestones of Britain, became extinct at the end of the Permian; those that evolved later (scleractinian types) have given rise to the modern corals like those of the Great Barrier Reef in Australia. Just as the Great Barrier Reef forms an imposing structure today,

Right:
Solitary rugose
corals from the
Carboniferous of
Derbyshire.

Far right:
Fenestella, a net-
like bryozoan from
the Carboniferous
Limestone of
North Wales.

so many fossil examples form the core of much of the limestones seen in the Carboniferous of northern England, and their Jurassic equivalent of the Midlands.

BRYOZOANS

These are unassuming colonial fossils that, in some cases, are confused with corals and, in other cases, appear as mat- or net-like fossils attached to the surface of others. Often overlooked by fossil collectors, bryozoans are actually complex organisms that have been around since the Ordovician, and, prior to the great Permian extinction event, helped form some of the largest reefs during the Palaeozoic.

CRABS, SHRIMPS AND LOBSTERS

These crustaceans are relatives of the trilobites, and have fragile external skeletons that fall apart when they moult, and after their death. Relatively rare because of their fragility, crustaceans have nonetheless been around since the Cambrian, often identified by the burrows that are commonly seen in sedimentary rocks.

A fossil crab
preserved in the
Eocene London
Clay of the Isle of
Sheppey in Kent.

CRINOIDS

These are common fossils in Palaeozoic rocks, so common in fact that whole bands of limestone are made up of their remains, particularly in the Carboniferous of northern England. Sometimes confusingly named 'sea lilies', because of their resemblance to stalked flowers when whole, these organisms are actually related to starfish and sea urchins, part of the major group called the echinoderms. Crinoids have stalks made up of many plates (called ossicles), which fell apart easily on death. On top of the stalk was an enclosed cup (which resembles, in part, the shell of its relative, the sea urchin), and a set of arms to gather food. Crinoids were a major part of Palaeozoic reefs, but were to be cut back in numbers at the end of the

A crinoid, complete with stem, cup and arms, preserved in the Carboniferous Limestone of southern Ireland.

Permian; there are relatively few species existing in today's oceans. Their relatives, the starfish and brittle stars, were equally fragile, but around in fewer numbers; they are even rarer.

SEA URCHINS

These are commonest from the Mesozoic onwards, and consist of a box (called the test) that is made up of numerous plates, each one a crystal of the mineral calcite (calcium carbonate). In life, sea urchins are covered with a thin skin covered with spines of varying sizes (which gives them their group name, echinoderm) as anyone who has been on holiday to the Mediterranean will attest – grave warnings are given out to small children to avoid stepping

Below left: *Hemicidaris*, a modern preserved specimen of a regular sea urchin, complete with spines.

Below: *Echinocardium*, a modern preserved specimen of a burrowing 'heart urchin'.

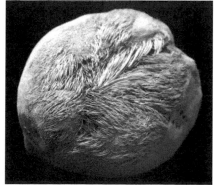

on dark green spiky objects lurking in the rocks. In the main, sea urchins that live on rocky surfaces have regular domed shells with large spines. Burrowing versions are typically heart-shaped (allowing them to burrow more effectively), and have more subdued spines. Sea urchins are particularly common in the Jurassic limestones of the Cotswold Hills, and in the Cretaceous Chalk of south-east England.

CLAMS

These are the familiar seashells we have all collected from beaches at some time in our lives. Actually consisting of two shells (known as valves) that hinge together, clams are also known as bivalves, and there is a great diversity of types. Burrowing clams are usually symmetrical, the two valves mirroring each other. Others, such as oysters and scallops, often have valves of differing shapes and sizes (the exception being mussel shells), living on the surface. Commonly, the two valves fall apart when the animal dies, as they are held together by muscles acting against a spring-like ligament,. Bivalves have been around for a long time, first appearing in the Ordovician, but they really hit the big time after the widespread decimation of the brachiopods at the end of the Permian, and are commonest from the Triassic onwards. They are part of the Mollusca – barring the insects, one of the most diverse groups of invertebrate organisms. Clams are particularly common, starring alongside their cousins, the sea snails, from the Palaeogene onwards.

Below:
Modern-day clams, showing the variety of types.

Below right:
Cardinia, clams (bivalves) from the Lower Jurassic of Yorkshire.

SEA SNAILS

These are amongst the most attractive of all modern seashells, with a diversity of shapes (though all based on the simple coiled cone) and surface colours. Fossil sea snails – technically known as gastropods (literally meaning 'mobile stomach'), have a similar diversity of shapes, but have largely lost their colour during the process of fossilisation. Although present in much

older rocks, gastropods are common from the Mesozoic onwards, and are significant fossils in rocks from the Palaeogene and Neogene (the Tertiary of older books), sitting alongside their contemporaries, the bivalves.

Above left: Modern-day sea snails, showing their variety of form.

AMMONITES

These are amongst the most attractive and prized of all fossils, the pleasing shape of their spiral form, the beauty of their ribs and internal structures providing an aesthetic feast for the eye. Unlike other molluscs, such as clams and seashells, ammonites are extinct, having died out alongside fellow victims the dinosaurs and marine reptiles during the end-Cretaceous extinction event. Ammonites were predators, similar in many ways to the 'living fossil' nautilus, with a well-developed nervous system, eyes and grasping arms. Floating using their buoyant, chambered, shells, like nautilus they had an ability to swim backwards using a kind of water 'jet-propulsion' system. Ammonites are prized by geologists, as they evolved rapidly and were widespread; as such they are useful tools in matching rocks by age across the world. Ammonites and their similar ancestors first appeared in the Devonian, and died out at the end of the Mesozoic.

Above: *Hildoceras bifrons*, an ammonite from the Lower Jurassic of Whitby. This ammonite is named after St Hilda, who is said to have turned all serpents to stone, and cast them into the sea.

BELEMNITES

These are mostly bullet-shaped fossils that lived alongside their ammonite cousins. The belemnite shell was internal, and has a chambered part, and an object known as the 'guard', which acted as a counterbalance to keep its squid-like owner horizontal in the water column, and thus more ready to hunt using its arms. Belemnites used an ink sac like that of their squid cousins to evade attack. They are common fossils in Jurassic and Cretaceous rocks, but, like the ammonites, did not survive the end-Cretaceous extinction event.

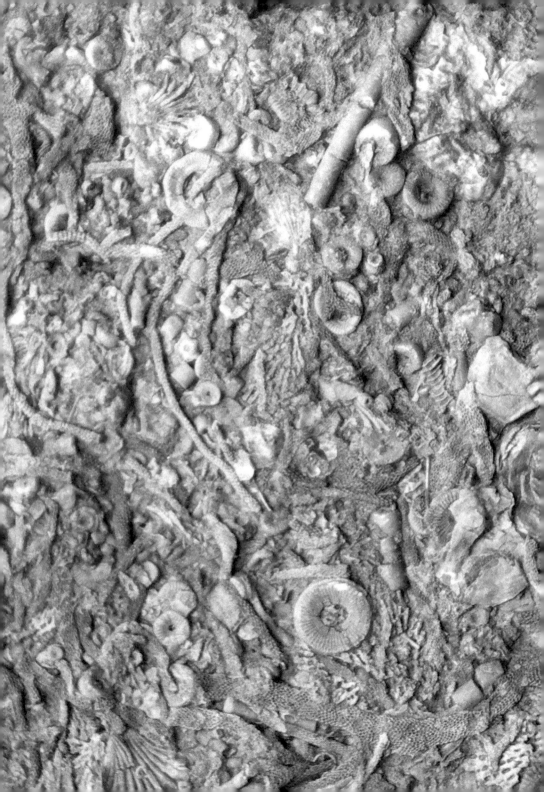

BRITAIN'S FOSSILS

Bᴿ RITAIN's fossils are rich and varied. This chapter provides a visual guide
to some of the commonest examples, grouped in the main divisions of
geological time. Precambrian fossils, present in Britain but few and far
between, are not illustrated. Each fossil is accompanied by its genus and
species name (where unidentified, denoted by the letters sp. for species),
which is usually given in italics with the author of the original species
afterwards. Where the genus has been changed by later discoveries, the
author's name is given in parentheses. The maximum dimension is given as
a guide.

LOWER PALAEOZOIC (CAMBRIAN—SILURIAN)

Cambrian rocks are found in the heart of Snowdonia, and it is from here that,
rarely, fossils are found. These include the trilobite *Paradoxides*, which is also found
in parts of Europe, but not Scotland, helping to demonstrate the existence of a
long-dead ocean, Iapetus, that once separated these parts of Britain. Cambrian
fossils are less common than those of the Ordovician and, particularly, the Silurian
rocks of Wales, the Welsh borderland, and parts of the English Midlands.

Opposite:
Bryozoans and
crinoids from the
Silurian of the
Welsh borderland.

Trinucleus fimbriatus Murchison
Trilobite, partial specimen, 19 mm
Ordovician (Caradoc), Builth Wells.

Ogyiocarella debuchii
Trilobite, specimen with partial head, 40 mm
Ordovician (Caradoc), Builth Wells.

Dalmanites myops (König)
Trilobite, heads and solitary tail, 25 mm
Silurian (Wenlock), Upper Milichope, Shropshire.

Calymene blumenbachi Brongniart
Trilobite, head with accompanying bryozoan, 28 mm
Silurian (Wenlock Limestone), Dudley, West
Midlands.

Didymograptus murchisoni (Beck)
Graptolite, largest specimen 48 mm
Ordovician (Llanvirn), Abereiddy Bay, Pembrokeshire.

Climacograptus wilsoni Lapworth
Graptolite, 24 mm
Ordovician (Llandovery), Dobbs Linn, Moffat.

Monograptus sp.
Graptolite, 21 mm
Silurian (Wenlock), Upper Milichope, Shropshire.

Lingulella davisii (M'Coy)
Brachiopod, 20 mm
Cambrian (Tremadoc), Porthmadoc, North Wales.

Dalmanella horderleyensis (Whittington)
Brachiopod, internal mould, 25 mm
Ordovician (Caradoc), Shropshire.

Leptaena depressa (Sowerby)
Brachiopod, 21 mm
Silurian (Wenlock Limestone), Dudley, West
Midlands.

Strophonella euglypha (Dalman)
Brachiopod, 41 mm
Silurian (Wenlock Limestone), Dudley, West
Midlands.

Atrypa reticularis (Linnaeus)
Brachiopod, three specimens, 12 mm
Silurian (Wenlock Limestone), Dudley, West
Midlands.

Favosites gothlandicus Lamarck
Coral (tabulate), two small examples, 45 mm
Silurian (Wenlock Limestone), Wenlock Edge,
Shropshire.

Halysites catenularius (Linné)
Coral (tabulate), colony, 58 mm
Silurian (Wenlock Limestone), Wenlock Edge,
Shropshire.

Bryozoans and crinoids, block size 90 mm
Silurian (Wenlock Limestone), Wenlock Edge,
Shropshire.

Poleumita discors (Sowerby)
Gastropod, diameter 58 mm
Silurian (Wenlock Limestone), Wenlock Edge,
Shropshire.

UPPER PALAEOZOIC (DEVONIAN, CARBONIFEROUS, PERMIAN)

The Upper Palaeozoic rocks of Britain contain a rich assemblage of fossils, none more so than the Carboniferous, which contains a range of marine fossils in the limestones that form the Pennine Hills, parts of North Wales, much of the Midland Valley of Scotland, and much of Ireland. Devonian rocks in their home county also contain reef fossils, contrasting with the largely land-based Old Red Sandstone seen elsewhere. Fossil fish from Caithness are its most famous fauna. The Permian limestones of northern England also contain some spectacular fossils.

Cyrtospirifer extensus (Sowerby)
Brachiopod ('Delabole butterfly'), deformed specimen in slate, 70 mm
Upper Devonian, Delabole, Launceston.

Productus sp.
Brachiopods, largest 32 mm
Lower Carboniferous (Viséan, Carboniferous Limestone), Halkyn Mountain, Clwyd.

Gigantoproductus giganteus (Sowerby)
Brachiopod, 130 mm
Lower Carboniferous (Viséan, Carboniferous Limestone), Trefor Rocks, Llangollen.

Antiquatonia hindi (Muir-Wood)
Brachiopod, crushed but preserved with spines intact, 43 mm
Lower Carboniferous (Viséan, Carboniferous Limestone), Trefor Rocks, Llangollen.

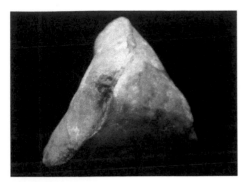

Pugnax acuminatus Sowerby
Brachiopod, 32 mm
Lower Carboniferous (Viséan, Carboniferous
Limestone), Derbyshire.

Spirifer striatus (Martin)
Brachiopods, 28 mm
Lower Carboniferous (Viséan, Carboniferous
Limestone), Halkyn Mountain, Clwyd.

Polypora sp.
Coral (tabulate, with rugose example mid-left),
polished section, 50 mm
Devonian, Torquay, Devon.

Lithostrotion portlocki (Bronn)
Coral (rugose, compound), sliced and polished
section, 52 mm
Lower Carboniferous (Viséan, Carboniferous
Limestone), Derbyshire.

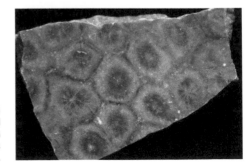

Palaeosmilia regium (Phillips)
Coral (rugose, compound), sliced and polished
section, block size 77 mm
Lower Carboniferous (Viséan, Carboniferous
Limestone), Northumberland.

Dibunophyllum bipartium (M'Coy)
Coral (rugose, solitary), 'horn coral', 29 mm
Lower Carboniferous (Viséan, Carboniferous
Limestone), Derbyshire.

Dibunophyllum bipartium (M'Coy)
Coral (rugose, solitary), sliced and polished section,
38 mm diameter
Lower Carboniferous (Viséan, Carboniferous
Limestone), Derbyshire.

Dipteris valenciennes
Fish, posterior portion, 200 mm length
Devonian (Caithness Flags), Spittal, Caithness.

Crinoid ossicles, 18 mm maximum diameter
Lower Carboniferous (Viséan, Carboniferous
Limestone), Derbyshire.

Gastrioceras carbonarium (Buch)
Ammonite (goniatite), pyritised example,
58 mm diameter
Upper Carboniferous (marine bands, Coal
Measures), Halifax, North Yorkshire.

Carbonicola communis Davis and Trueman
Freshwater bivalves, 31 mm
Upper Carboniferous (Lower Coal Measures),
Wakefield.

Lepidodendron sp.
Plant ('club moss'), section showing leaf scars,
110 mm
Upper Carboniferous (Coal Measures), Northern
England.

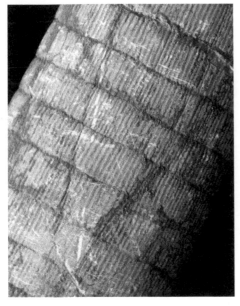

Alethopteris decurrens (Artis)
Plant (fern), part of frond, 53 mm
Upper Carboniferous (Lower Coal Measures),
Dowglais near Merthyr Tydfil.

Calamites sp.
Plant (horsetail), part of stem, 105 mm length
Upper Carboniferous (Coal Measures), Blisthorpe
Colliery, Newark, Nottinghamshire.

MESOZOIC (TRIASSIC, JURASSIC, CRETACEOUS)

Fossils are found in abundance in the Mesozoic rocks of the UK, from Northern Ireland, parts of Scotland, and particularly south-east of the Tees–Exe line that runs across England. Sitting on the largely unproductive red sandstones of the Triassic, Jurassic and Cretaceous limestones, shales and clays contain some of the most beautiful fossils to be collected in Britain.

Psiloceras planorbis (Sowerby)
Ammonite, crushed example (usual preservation), 43 mm diameter
Lower Jurassic (Lower Lias), Watchet, Somerset.

Asteroceras obtusum (Sowerby)
Ammonite, calcite-filled chambers, 20 mm diameter
Lower Jurassic (Lower Lias), Charmouth, Dorset.

Liparoceras cheltiense (Murchison)
Ammonite, with original shell preservation, 51 mm
Lower Jurassic (Lower Lias), Blockley, Gloucestershire.

Dactylioceras tenuicostatum (Young and Bird)
Ammonite, 70 mm
Lower Jurassic (Upper Lias), Kettleness, North Yorkshire.

Kosmoceras pollux (Reinecke)
Ammonite, usual flattened preservation, 42 mm
Middle Jurassic (Oxford Clay), Dorset.

Pavlovia pallasioides (Neaverson)
Ammonite, usual flattened preservation, 40 mm
Upper Jurassic (Kimmeridge Clay), Dorset.

Anahoplites planus (Mantell)
Ammonite, pyritised, 25 mm
Lower Cretaceous (Gault Clay), Folkestone, Kent.

Aegiocrioceras quadratum (Crick) (top) and
Endemoceras regale (Pavlov) (bottom)
Ammonites, with original shell material, largest
Endemoceras 39 mm; *Aegiocrioceras*, 41 mm, is an
'unrolled' (known as heteromorph) ammonite
Lower Cretaceous (Speeton Clay), Filey, North
Yorkshire.

Turrilites acutus Passy
A vertically spiralled (heteromorph) ammonite, 48
mm length
Upper Cretaceous (Lower Chalk), Dorset.

Acrocoelites vulgaris (Young and Bird) (left) and *A.*
subtenuis (Phillips) (right)
Belemnites, 70 mm
Lower Jurassic (Upper Lias), Saltwick, North
Yorkshire.

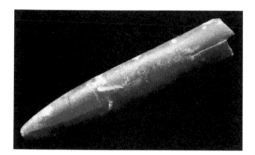

Belemnitella mucronata (Schlotheim)
Belemnite, 87 mm
Upper Cretaceous (Upper Chalk), Thanet, Kent.

Pleuromya costata (Young and Bird)
Bivalve, 38 mm
Lower Jurassic (Lower Lias), Blockley,
Gloucestershire..

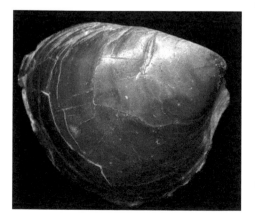

Plagiostoma giganteum Sowerby
Bivalve, 66 mm
Lower Jurassic (Lower Lias), Dorset.

Gryphaea arcuata Lamarck
Bivalves, 70 mm
Lower Jurassic (Lower Lias), Lavernock Point, South
Wales.

Myophorella clavellata (Sowerby)
Bivalve, 57 mm
Middle Jurassic ('Coral Rag', Oxfordian), Weymouth.

Spondylus spinosus (Sowerby)
Bivalve, 46 mm
Upper Cretaceous (Upper Chalk), Huntingdon.

Thamnastria sp.
Coral (scleractinian), 'Tisbury Star Stone', silicified,
70 mm
Upper Jurassic (Portlandian), Tisbury,
Gloucestershire.

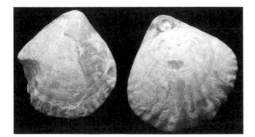

Plectothyris fimbriata (Sowerby)
Brachiopod, 21 mm
Middle Jurassic (Inferior Oolite), nr. Stroud,
Gloucestershire.

Moutonithyris dutempleana (d'Orbigny)
Brachiopod, 29 mm
Lower Cretaceous (Red Chalk), Hunstanton,
Norfolk.

Hemicidaris intermedia (Fleming)
Sea urchin (echinoid), crushed, 32 mm
Middle Jurassic (Coralline Oolite), Calne, Wiltshire.

Clypeus ploti (Salter)
Sea urchin (echinoid), 53 mm
Middle Jurassic (Inferior Oolite), Gloucestershire.

Echinocorys scutatum Leske
Sea urchin (echinoid), flint internal mould, 63 mm
Upper Cretaceous, Thanet, Kent.

Pentacrinites fossilis Blumenbach
Crinoid, part of the arms and cup, 69 mm
Lower Jurassic (Lower Lias), Charmouth, Dorset.

Micraster coranguinum (Leske)
Sea urchin (echinoid), 'heart urchin', 60 mm
Upper Cretaceous, Thanet, Kent.

Dinosaur (Iguanodon?) bone, 40 mm
Lower Cretaceous, Atherfield, Isle of Wight.

Meyeria sp.
Crustacean, 51 mm
Upper Greensand, Atherfield, Isle of Wight.

Mosasaur (marine reptile) teeth, 50 mm max.,
Lower Cretaceous (Cambridge Greensand),
Cambridge.

CENOZOIC (PALAEOGENE, NEOGENE, QUATERNARY)

The soft sediments of the Cenozoic, found in abundance around the coasts of southern Britain, are constantly being eroded by the actions of the sea. This is uncomfortable for those who live there, but fortuitous for fossil collectors, who can bank on fresh exposures of clay and other sediments rich in the organic remains of former worlds. The Isle of Wight and the coasts of Hampshire, as well as Essex and other parts of East Anglia, are rich in fossils.

Brotia melanoides (Sowerby)
Gastropod, 50 mm
Palaeogene (Eocene, Woolwich Beds), Charlton, south-east London.

Clavilithes macropira Cossman
Gastropod, 77 mm
Palaeogene (Eocene, Barton Beds), Barton, Hampshire.

Sassia arguta (Solander)
Gastropod, 52 mm
Palaeogene (Eocene, Barton Beds), Barton, Hampshire.

Athleta athleta (Solander)
Gastropod, 48 mm
Palaeogene (Eocene, Barton Beds), Barton, Hampshire.

Neptunia contraria (Linné)
Gastropod, left-hand coiling whelk, 75 mm
Quaternary (Red Crag), Walton-on-the-Naze, Essex.

Viviparis angulosus (Sowerby)
Freshwater gastropod, 26 mm
Palaeogene (Eocene, Bembridge Limestone),
Whitecliff Bay, Isle of Wight.

Natica multipunctata Wood
Carnivorous gastropods, 23 mm. The borehole was
made by another example of the same species.
Quaternary (Red Crag), Walton-on-the-Naze,
Essex.

Polymesoda cordata (Morris)
Bivalves, 30 mm
Palaeogene (Eocene, Woolwich Beds), Charlton,
south-east London.

Venericor planicosta (Lamarck)
Bivalves, 54 mm
Palaeogene (Eocene, Bracklesham Beds),
Bracklesham, Sussex.

Pholadomya margaritacea (Sowerby)
Bivalve, 55 mm
Palaeogene (Eocene, London Clay), Alum Bay, Isle of
Wight.

Glycimeris glycimeris (Linné)
Bivalves, 37 mm
Quaternary (Red Crag), Walton-on-the-Naze, Essex.

Glycimeris brevirostris (Sowerby)
Bivalve, 44 mm
Palaeogene (Eocene, London Clay), Bognor Regis,
Sussex.

Sharks' teeth, 5 mm
Palaeogene (Eocene, Blackheath Beds), Abbey Wood,
south-east London.

Aquipecten operculis (Linné)
Bivalves, largest 22 mm
Neogene (Pliocene, Coralline Crag), Sutton, Suffolk.

Whale bone fragment, 70 mm
Quaternary (Red Crag), Walton-on-the-Naze,
Essex.

COLLECTING FOSSILS

FOSSILS from Britain have been studied for hundreds of years. Some are easy to find, others more challenging. Knowing where to collect, what to collect, and when it is not possible to collect is important. Today, with increased pressure on our natural environment, it is important to act responsibly, but there is no reason why you should not be able to build up a representative collection of British fossils. This chapter guides you through the approaches to collecting your own fossils.

The study of fossils and ancient life – palaeontology – is accessible to all, as it is possible to make a collection of fossils that is representative of the ancient inhabitants of the seas that once covered Britain, evidence of ancient 'greenhouse' periods long before human-controlled atmospheric emissions. Carefully collected, and properly curated at home, such collections can form the basis of a much deeper understanding of extinct organisms, as well as providing an absorbing and fascinating hobby. People have been aware of fossils as cultural artefacts for centuries, millennia even; they have been used as decorative objects, objects to ward off evil and illness, and even as playthings. But collecting fossils for study has been a pastime since the birth of the science of palaeontology in the last part of the eighteenth century, and many of the UK's museums contain collections donated by private individuals. Sadly, the value of some of these collections is undermined by the fact that many of the finds were poorly recorded on collection, or the labels were subsequently lost. Lacking records, fossils become little more than decorative objects, and many of the commonest examples for sale in fossil shops (now mostly imported) come with the barest of details. This need not be the case. Today, fossil-collecting codes and guidelines have been published by the conservation agencies, by museums and clubs, and by responsible fossil-collecting websites. These guides ensure that the hobby of fossil collecting has a long-term future, and that significant finds are saved for scientific study in museums and universities. A simplified fossil code, based on this advice, is given here (page 62).

One boy's haul of fossils from Walton-on-the-Naze, Essex.

Opposite:
One drawer, well curated, of a fossil collection, with each fossil labelled and housed in its own box.

A Victorian 'cabinet of curiosities', filled with fossils, a collection that still holds its intrinsic value today.

The trilobite *Flexicalymene*, bought at a flea market in southern Spain. Devoid of information, this Moroccan fossil is little more than a decorative object.

Searching the beach for fossils washed from the cliffs in Dorset. Wave-cut rocky foreshores are likely to be more productive than pebble beaches.

PRACTICAL GUIDELINES

In setting out to collect fossils, it is important to remember that cliffs, rock faces and other types of unstable overhanging rocks are a danger to life and limb. Coastal rock exposures often seem like attractive places for collecting fossils, as the action of the sea constantly refreshes rock exposures by eroding them back – thereby supplying fresh specimens that can be collected from the

Coastal cliffs at Charmouth, Dorset, looking towards Golden Cap. Part of the 'Jurassic Coast' World Heritage Site, a fossil-collecting code directs would-be collectors away from the dangerous cliffs, and on to the foreshore.

Above:
Fossil-collecting
equipment.
Hammers are not
required in all
situations, and
should be wielded
with care,
particularly on
harder rocks.
Goggles are
advisable. More
important are
boxes, bags,
notebooks and
other means of
recording and
protecting the
details of your finds.

Above right:
Probes and mini
power tools like
this one, supplied
with a range of
heads and bits, are
useful in preparing
fossils. Remember
to wear goggles and
a dust mask.

beach. But much care must be exercised in collecting from beaches and coastal rock outcrops. In some places, especially those visited regularly by fossil collectors, such as Lyme Regis or Charmouth on the Dorset coast, it may be difficult to find new fossils at peak times. (However, the activity of the sea is such that fresh samples can be obtained after stormy intervals.) Do not be tempted to hammer the cliffs in order to get your own, fresh samples; often they will be overhanging, unstable and dangerous. The same is true in all other situations where there is a vertical rock face, such as in an old quarry, an inland cliff, or a road cutting. On the coast, it is also essential to keep a watch on rising tides; rocky coasts are usually treacherous in this respect, and many unwary collectors have had to be rescued.

Fossils are common in some rocks, and often the most attractive fossils are those that have been eroded out from the host rocks. In the past, hammers and chisels have been seen as essential tools in order to assemble a collection. But today, most authorities would advise that hammers should not routinely be used. Stone chips produced when hammering some types of rock, such as limestone, often fly off from the point of impact, and invariably head towards the eyes of the hammerer, or the observer. If hammers are to be used, then safety goggles are essential. It is as well to remember that geological hammers are made from especially hard, tempered steel and are more robust than the average domestic hammer; using a domestic hammer on hard rocks can lead to shards of steel being chipped off its head, which could, like the stone chips, easily end up in the eyes.

Extracting fossils from their enclosing rocks is a difficult business and is no mean feat when it comes to hard rocks such as limestones, which are brittle and tend to break into shards when hit with a hammer. This is as good a reason as any to persuade you to collect only loose specimens, wherever possible. However, where this is not possible, then careful trimming of excessive rock with a hammer, followed by the use of steel probes and needles to work carefully around the specimen, is a useful technique where fossils are found in hard rocks. It is often difficult to extract some fossils from their enclosing rock, particularly limestones, and hard, well-cemented sandstones. Small engraving tools, cheaply available and with a range of heads and bits, are valuable in preparing specimens found in harder rocks (remembering the need for goggles and a dust mask). It is usual to leave the fossil partially enclosed within a small block of matrix, and in so doing, work around the fossil to make it stand proud of a smooth surface. In softer rocks, such as soft sandstones, mudstones and clays, fossils can often be removed whole with a bit of care and patience, but some specimens, often those with spines and other delicate features, can be extremely fragile, so take considerable care when trying to remove them, as they can just as easily break or disintegrate. In such cases, retaining some of the original matrix as a support is usual. After preparation, display is really a matter of choice. Selection of a suitable set of drawers, with labelled boxes, and a list of contents is usually the first step, but many larger examples such as larger ammonites, for example, rightly end up as mantelpiece specimens.

Fossil-collecting code	
1	Always seek permission to collect from private land.
2	Always be aware of your safety, and of the safety of others. Wear appropriate safety equipment (goggles, hard hats).
3	Don't be tempted to over-collect. Leave some specimens for others.
4	Don't hammer for the sake of it, and don't ruin rock outcrops for others.
5	Collect loose specimens wherever possible, and take adequate packing materials to wrap them for the journey home.
6	Always record details of your find in the field. This should include: exact location, a description of the occurrence (including details of whether the fossils were found with others, how they were preserved and so on).
7	Be aware of the scientific value of your finds. If you think they are unusual, take them to your local museum for identification. Never discard your collection carelessly: offer it to others if you wish to dispose of it.
8	Curate your collection properly, keeping records of the finds, and storing them in appropriate boxes, trays and drawers.

An Edwardian approach to curation: labelled matchboxes have provided a home for these fossil brachiopods for the last hundred years.

Below right: A trilobite from Devonian rocks of Djebel Issoumour, Morocco. Attractive fossils such as this one may be obtained from fossil shops across the world.

Below: A Madagascan ammonite (a perisphinctid), beautifully preserved. Examples like this one often lack detailed information when bought from fossil shops.

CURATING YOUR COLLECTION

Collecting fossils is a rewarding pastime, but it does carry some responsibilities. You could be the first person ever to find a particular fossil species and, as such, it is incumbent upon you to collect specimens with care,

and treat them respectfully, as you would any collection of precious objects. The principal responsibilities are in the recording of the find itself – where the specimen came from, when it was collected, whether it was loose, or in place (*in situ*) in the rocks. If collected *in situ*, you should describe the layer from which it was obtained, ideally making a record of the find using sketches and photographs. All of these records will be of the utmost importance to anyone who might have the remotest scientific interest in a geological location. Remember all those museum specimens with no details? Perhaps your specimens will end up in a museum and, if so, they will have the greatest value if they are complete with details. Once obtained, it is best to keep fossils in small boxes or trays, so that they do not roll around inside drawers and become broken and damaged. Labels can be affixed to the specimens themselves, or to the boxes or trays, and a catalogue kept – these aspects are all part of responsible collecting. It is up to you how far you wish to engage in the hobby.

In the past it was usual to attach paper labels to fossils, recording the details of the find, and these have the benefit of being hardy. Numbers attached to the fossil, related to an explanatory key, are another method of ensuring valuable information is not lost.

BUYING FOSSILS

If you are unable to collect directly yourself, then reputable fossil dealers can provide some of your needs. Many shops are to be found in fossil-rich areas of the UK, with famous examples in the Lyme Regis and Whitby districts of England – and there are many more besides. Several will trade on the internet. Buying fossils is an easy way of building a collection, but, if you want anything more than a simple decorative object, you should insist on the fullest information being supplied with the specimen (itself collected according to appropriate codes), so that your collection will be of equivalent scientific value to one you may have collected yourself *in situ*. It is important to remember that fossil shops and internet sites will have a stock derived mostly from overseas sources – trilobites from Morocco, ammonites from Madagascar, for example – but if you want a collection of British fossils, you will have to shop around.

SOURCES OF INFORMATION

Local museums are good sources of information to help you with your collection, and will often have specimens on display. Such museums may have

These three handbooks have provided guidance for all collectors of British fossils for over forty years. Sadly, they are now out of print, but are commonly found in second-hand bookshops.

a geological curator who might be able to advise, and who will undoubtedly be interested in your fossil if it is a new find, thereby adding to scientific knowledge. The internet has greatly expanded the possibilities of research for fossil collectors – but do not underestimate the value of books in this field. Surprisingly, still the best guides to British fossils remain those first published by the Natural History Museum (then the British Museum [Natural History]) in the 1960s. Packed with detailed line drawings, these three books (*British Palaeozoic Fossils*, *British Mesozoic Fossils*, and *British Caenozoic Fossils*) still provide some of the fullest information available. Sadly out of print, they can nevertheless be easily and cheaply obtained from second-hand booksellers. A more advanced series of guides, published by the Palaeontological Association and written by experts in the field, provides some advanced guidance for specific rock units, as do the many fossil websites that are starting to proliferate.

SUGGESTED READING

GENERAL BOOKS

Doyle, Peter (1996) *Understanding Fossils*. John Wiley, Chichester.

Oakley, Kenneth P. (1985) *Decorative and symbolic uses of fossils*. Pitt Rivers Museum, Oxford.

Scottish National Heritage (2008) *Scottish Fossil Code*. SNH, Edinburgh.

Thackray, John (1984) *British Fossils*. HMSO, London.

Thompson, Keith (2005) *Fossils. A Very Short Introduction*. Oxford University Press, Oxford.

Walker, Cyril and Ward, David (1992) *Fossils*. Eyewitness Handbooks, Dorling Kindersley, London.

IDENTIFICATION GUIDES

British Museum (Natural History) (1975) *British Caenozoic Fossils*. Fourth Edition. BMNH, London.

British Museum (Natural History) (1975) *British Palaeozoic Fossils*. Fourth Edition. BMNH, London.

British Museum (Natural History) (1983) *British Mesozoic Fossils*. Sixth Edition. BMNH, London.

Cleal, Chris J. and Thomas, Barry J. (1994) *Plant Fossils of the British Coal Measures*. Palaeontological Association, London.

Harper, David A.T. and Owen, Alan W. (eds) (1996) *Fossils of the Upper Ordovician*. Palaeontological Association, London.

Martill, David M. and Hudson, John D. (eds) (1991) *Fossils of the Oxford Clay*. Palaeontological Association, London.

Smith, Andrew B. and Batten, David J. (eds) (2002) *Fossils of the Chalk*. Second Edition. Palaeontological Association, London.

WEBSITES

www.discoveringfossils.co.uk and www.ukfossils.co.uk are both useful guides.

www.rockwatch.org.uk (Young Persons' Geology Club)

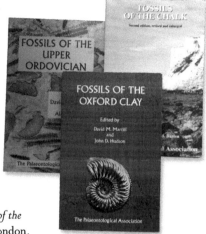

The Palaeontological Association produces high-quality guides to the fossils of several classic British fossil-bearing rock units; more are in production.

INDEX